YOUR KNOWLEDGE HAS VALUE

Bradley Tice

Algorithmic Complexity and Plant Genetics

GRIN Verlag

Bibliografische Information der Deutschen Nationalbibliothek:

Die Deutsche Bibliothek verzeichnet diese Publikation in der Deutschen National-
bibliografie; detaillierte bibliografische Daten sind im Internet über http://dnb.d-
nb.de/ abrufbar.

Imprint:

Copyright © 2014 GRIN Verlag GmbH
Druck und Bindung: Books on Demand GmbH, Norderstedt Germany
ISBN: 978-3-656-59146-7

This book at GRIN:

http://www.grin.com/en/e-book/268096/algorithmic-complexity-and-plant-genetics

GRIN - Your knowledge has value

Der GRIN Verlag publiziert seit 1998 wissenschaftliche Arbeiten von Studenten, Hochschullehrern und anderen Akademikern als eBook und gedrucktes Buch. Die Verlagswebsite www.grin.com ist die ideale Plattform zur Veröffentlichung von Hausarbeiten, Abschlussarbeiten, wissenschaftlichen Aufsätzen, Dissertationen und Fachbüchern.

Visit us on the internet:

http://www.grin.com/

http://www.facebook.com/grincom

http://www.twitter.com/grin_com

Algorithmic Complexity and Plant Genetics

Dr. Bradley S. Tice

ABSTRACT

The paper will present a compression program algorithm that will compress sequential strings of plant DNA and RNA for storage and transmission of plant genetic information. The need for compression of plant genetic data will be examined in both a theoretical manner and a practical manner in regards to large data pools of plant genetics information, Big Data, and genetic code space saving techniques for applied plant genetics.

Introduction

The paper will present an algorithm program that can compress random and non-random sequential strings that can be applied to plant genetics. Both plant DNA and RNA can be compressed from the original structures genomic length. This has a direct application for storage and transmission of large data pools of agriculturally important plant stock genetics information. The algorithm used for the compression and de-compression of plant genetic information was discovered by the author in 1998 and is the most accurate and precise measure of randomness known [1].

Compression Algorithm

The algorithm compression program uses the traditional left to right input of a segment of a sequential string of characters, in this case individual genetic DNA or RNA molecules, that is then sub-grouped into like natured characters, that can be compressed into a complete compression, a universal compression, or a 'specific' or 'partial' compression of the compressed sequential string [2].

In the compression of like natured genetic material on the original sequential string of a plant's genetic code the resulting plant's genetic code is reduced, compressed, without either the type or placement of that genetic information being lost. This has plant research and development applications to plant genetics as it allows for space saving techniques to theoretical genetics and practical applications to applied genetics research.

A Compression Algorithm: Some Examples

If a sequential string of binary characters represent a digital representation of a plants genetic code, a theoretical model of that code, representing a translation from the plants original analog genetic code sequence; an alpha symbol system, can be 'compressed' for storage and transmission purposes within a digital and computer communications network.

Example A: The following binary sequential string is a composite of a random sequential binary string.

Example A: [1111100010001110001110000010]

If the linear sequential string of [1] and [0] of Example A is separated into sequentially common sub-groups the following will result:

Sub-group of Example A: [11111] + [000]+[1]+[000]+[111]+[000]+[111]+[00000]+[1]+[0]

The non-random features of the sub-groups are those that have a 'pattern' to the sequence of [1] or [0] characters such as these sub-groups:

Non-random sub-groups: [000]+[111]+[000]+[111]

Each is a three character sub-grouping of either [1] or [0] and can be compressed as 0101 and notated as a sub-group of the initial character, either a [1] or a [0], of each sub-group composed of the same 3 characters total.

The remaining sub-groups are not as patterned' as the non-random sub-group and are referred to as random sub-group sequences of a sequential binary string.

The remaining random sub-group sequences are as follows:

Random sub-group: [11111]+[000]+[1]...[00000]+[1]+[0]

The random sub-group can be compressed by notating the number of total like natured digits, either [1] or [0], with a suffix number to denote the total number of characters following the initial digit.

[11111] = [1x5]

[000] = [0x3]

[1] = [1]

[00000] = [0x5]

[1]= [1]

[0] = [0]

The original sequential binary string of Example A was as follows:

Example A: 1111100010001110001110000010

The notated compressed form of Example A is as follows:

Notated Example A: 1x5 0x3 1 0x3 1x3 0x3 1x3 0x5 1 0

The non-notated compressed state is as follows:

Non-notated Example A: 1010101010

A compressed state of 10 characters from the original 28 character length.

Big Data

Because digital storage and transmission are used in large data set collections, the use of the traditional binary format of [0] and [1] are used to transcribe the analog world into the digital world of computing. The vast amounts of plant data makes the need for 'interpreting' that data into a comprehensible whole a growing need in the biological sciences [3]. Due to the rich diversity of both natural and engineered plants, the practical problems of gaining insight into all this plant genetic data to form some type of plant genetics 'information', information being the resulting product of 'work' obtained from the scholarly use of the intellectually 'neutral' plant genetic data, let alone the storage and transmission of such amounts of genetic data are overwhelming at best [4] & [5].

Plant Genetics

Even with large amounts of plant diversity, the growing need for new strains of 'food stock' plants has increased the scientific interest in the world's seed banks [6], [7] & [8]. Not just the amount of food needed for future consumption, but the quality and types of foods derived from modified and indigenous genetic mutations [9] & [10].

High value crops such as grapes and citrus are having their genome sequenced to find the 'biological' history of each strain and to develop a data base from which a genetic road map can be designed for the improvement of each type of plant [11], [12] & [13]. All of these developments within agriculture and biology necessitate large volumes of data that need to be stored and transmitted into a digital medium.

The result is an ever expanding amount of data that must be searched, stored, transmitted long before it can be analyzed for 'information' rich purposes. The algorithmic compression program presented in this research paper is fundamentally at the crossroads for such high level compression of large data base stores and the communication within a digital network.

Algorithmic Compression and Big Data

The continuous influx of plant data for analysis and storage has made the issue of 'Big Data' a primary area for computational biology to address in both techniques, software and hardware. The compression algorithm discussed in this paper is the most compressible system known and is applicable to both random and non-random forms of compression. The author is currently working on a large scale model of a complete 'design test' system of information processing, transmission and storage of data that will be the most advanced system for addressing the issue of large data pools in the biological sciences.

The results from these 'design tests' will be presented as papers at future conferences. The research work at Advanced Human Design is currently on the polymer sciences with an emphasis on plant genetics. Future papers on specific crop plant genetics are planned for the next few years.

Conclusion

The paper has presented research that confirms the application to the following aspects of theoretical plant genetics research:

- Theoretical plant genetics applications directly to models of a plants genetic code for both partial and whole compression, and de-compression, of sequentially common genetic types in the genetic line of a plants genetic code.

Applied aspects to plant genetics are as follows:

- Compression of large volumes of plant genetic data in the form of sequential lines of genetic information that can be compressed using both random and non-random genetic lines into a digital format and transmitted and stored in massive computer banks.

The theoretical use of an algorithmic compression program to plant genetics is a practical technique for compression of the plants genetic code and the applied aspect of a algorithmic compression program is that the analog chemical structures can be compressed, at points of adjoined sequences of the same type of character types, and transmitted and stored as 'compressed' linear sequential strings that can be 'de-compressed' without loss of number or character type.

3

The transitional aspect of algorithmic compression to chemical and structure notation systems from analog to digital format is a viable technique for compression of large amounts of sequential string data. The application to both theoretical and applied aspects of plant genetics is both universal and specific to each code sequence and has practical applications to other linear sequential string systems.

The practical application of this theory and applied model of algorithmic compression can be used on animal genetic codes with similar benefits to the analysis, transmission and storage of biological data. Again, the issue of large data sets of biological information can be accurately, rapidly and safely stored for future search use by scientists.

Summary

The algorithmic compression program uses the most accurate and precise measure of randomness known and applies this measure to a linear sequential string for maximal compression, and de-compression, of sequences data structures. The application to both theoretical and applied aspects to the biology of plant genetics is on many levels within the digital domain of modern scientific research and development.

With the advent of ever more powerful computers, the need for complimentary software systems will have to evolve with them as will the fundamental types of techniques used to analyze such large data pools of raw data. It is clear that the algorithmic compression system discussed in this paper has computer hard ware, software as well as transmission and storage application uses and the author is currently working on a 'design test' project at Advanced Human Design to refine these features in a theoretical computer information system

The paper has presented a strong case for the use of an algorithmic compression program for the vast amounts of biological data on plant genetics. Because plants are the most efficient food stock available and have the greatest likelihood for successful colonization of new strains and varieties, the application of bioinformatics and computational biology to the research on plant genetics should be a high priority for science and agriculture.

References

[1]. Tice, B.S. (2012) <u>A Level of Martin-Lof Randomness.</u>

New Hampshire: Science Publishers/CRC Press.

[2]. Tice, abide.

[3]. Cornell University (2013) "To feed the future: We must mine the wealth of the world's seed banks

today, experts argue". <u>Science Daily</u>, July 25, 2013, pp. 1-3.

Website: http://www.sciencedaily.com/releases/2013/07/13070512051

[4]. Crop Society of America (2013) "U.S. has surprisingly largest reservoir of crop plant diversity".

<u>Science Daily</u>, July 25, 2013, pp. 1-3.

Website: http://www.sciencedaily.com/eleases/2013/04/13042933536.htm

[5]. Nabhan, G.P. (2013) "Our coming food crisis". <u>The New York Times</u>, July 21, 2013, pp. 1-4.

Website: http://www.nytimes.com/2013/07/22/opinion/our-coming-food-crisis.html?hp&_r=0&page

[6]. Cornell University, abide.

[7]. Crop Society of America, abide.

[8]. Nabhan, abide.

[9]. American Academy of Microbiology (2013) "Report proposes microbiology's grand challenge to help feed the

world". <u>Science Daily</u>, September 2, 2013, pp. 1-2.

Website: http://www.sciencedaily.com/releases/2013/08/130872204536.htm

[10]. American Chemical Society (2013) "Science supporting abundant, nourishing food for a growing civilization".

<u>Phys.org</u>, September 8, 2013, pp. 1-2. Website: http://phys.org/print297880428.html

[11]. UNU-BIOLAC (2013) "Scientists sequence genome of high-value grape, seek secrets of wine's aroma". <u>Science</u>

<u>Daily</u>, September 2, 2013, pp. 1-3. Website:
http://www.sciencedaily.com/releases/2013/09/13090210184.htm

[12]. American Society for Horticultural Science (2011) "<u>Science Daily</u>, January 26, 2011, pp. 1-2.

Website: http://www.sciencedaily.com/releases/2011/01/110118101600.htm

[13]. Harmon, A. (2013) "A race to save the orange by altering its DNA". <u>The New York Times</u>, July 27, 2013,

pp. 1-14.

Website: http://www.nytimes.com/2013/07/28/science/a-race-to-save-the-orange-bty-altering-its-dna...